BOARD BULLETIN

The Assessment of Science Meets the Science of Assessment

Summary of a Workshop

Board on Testing and Assessment
Commission on Behavioral and Social Sciences and Education
National Research Council

National Academy Press
Washington, D.C.

NATIONAL ACADEMY PRESS 2101 Constitution Avenue, N.W. Washington, D.C. 20418

NOTICE: The project that is the subject of this report was approved by the Governing Board of the National Research Council, whose members are drawn from the councils of the National Academy of Sciences, the National Academy of Engineering, and the Institute of Medicine. The members of the committee responsible for the report were chosen for their special competences and with regard for appropriate balance.

The National Academy of Sciences is a private, nonprofit, self-perpetuating society of distinguished scholars engaged in scientific and engineering research, dedicated to the furtherance of science and technology and to their use for the general welfare. Upon the authority of the charter granted to it by the Congress in 1863, the Academy has a mandate that requires it to advise the federal government on scientific and technical matters. Dr. Bruce M. Alberts is president of the National Academy of Sciences.

The National Academy of Engineering was established in 1964, under the charter of the National Academy of Sciences, as a parallel organization of outstanding engineers. It is autonomous in its administration and in the selection of its members, sharing with the National Academy of Sciences the responsibility for advising the federal government. The National Academy of Engineering also sponsors engineering programs aimed at meeting national needs, encourages education and research, and recognizes the superior achievements of engineers. Dr. William A. Wulf is president of the National Academy of Engineering.

The Institute of Medicine was established in 1970 by the National Academy of Sciences to secure the services of eminent members of appropriate professions in the examination of policy matters pertaining to the health of the public. The Institute acts under the responsibility given to the National Academy of Sciences by its congressional charter to be an adviser to the federal government and, upon its own initiative, to identify issues of medical care, research, and education. Dr. Kenneth I. Shine is president of the Institute of Medicine.

The National Research Council was organized by the National Academy of Sciences in 1916 to associate the broad community of science and technology with the Academy's purposes of furthering knowledge and advising the federal government. Functioning in accordance with general policies determined by the Academy, the Council has become the principal operating agency of both the National Academy of Sciences and the National Academy of Engineering in providing services to the government, the public, and the scientific and engineering communities. The Council is administered jointly by both Academies and the Institute of Medicine. Dr. Bruce M. Alberts and Dr. William A. Wulf are chairman and vice chairman, respectively, of the National Research Council.

The study was supported by contract number ESI-935-5774 between the National Academy of Sciences and the National Science Foundation. Any opinions, findings, conclusions, or recommendations expressed in this publication are those of the author(s) and do not necessarily reflect the view of the organizations or agencies that provided support for this project.

International Standard Book Number 0-309-06546-1

Additional copies of this report are available from National Academy Press, 2101 Constitution Avenue, N.W., Washington, D.C. 20418

Call (800) 624-6242 or (202) 334-3313 (in the Washington metropolitan area)

This report is also available on line at **http://www.nap.edu**

Printed in the United States of America

BOARD ON TESTING AND ASSESSMENT

Membership as of the dates of the workshop summarized in this report.

ACKNOWLEDGMENTS

A great many people contributed to the success of this workshop, which brought together experts on science content and pedagogy and experts in educational measurement and evaluation. The Board would like to thank the presenters and discussants for their contributions to a lively and productive conference (see the agenda in Appendix A for complete list). In addition, we would like to acknowledge the contributions of the researchers who displayed posters and presented demonstrations of their work, which enriched the whole workshop. We are very grateful to the National Science Foundation, which, through the Education and Human Resources Directorate, has been an enthusiastic supporter of the Board's work and of this conference in particular.

Special thanks are due to board members Paul Black and Richard Shavelson for their help in conceiving, planning, and chairing this workshop. Staff member Kathy Guidroz deserves recognition for her efforts; thanks also to Senta Raizen and Nancy Carson, who served as consultants to the project. A number of staff contributed to the writing of this summary report, including Marilyn Dabady, Patricia Morison, and Naomi Chudowsky. We also wish to thank Nancy Kober and Robert Crossgrove for editorial help. Board director Michael Feuer offered his guidance and leadership to the project. Jan Liverance, Kimberly Saldin, and Viola Horek provided excellent administrative support.

This report has been reviewed in draft form by individuals chosen for their diverse perspectives and technical expertise, in accordance with procedures approved by the Report Review Committee of the National Research Council. The purpose of this independent review is to provide candid and critical comments that will assist the institution in making the published report as sound as possible and to ensure that the report meets institutional standards for objectivity, evidence, and responsiveness to the study charge. The review comments and draft manuscript remain confidential to protect the integrity of the deliberative process.

We wish to thank the following individuals for their participation in the review of this report: Joan Herman, National Center for Research on Evaluation, Standards, and Student Testing and Center for the Study of Evaluation, University of California, Los Angeles; Alan Lesgold, Learning Research and Development Center, University of Pittsburgh; Michael E. Martinez, Department of Education, University of California, Irvine; Leona

Schauble, Department of Educational Psychology, University of Wisconsin; and James Shymansky, Regional Institute for Science Education, University of Missouri, St. Louis.

While the individuals listed above have provided many constructive comments and suggestions, it must be emphasized that responsibility for the final content of this report rests solely with the authoring board and the institution.

CONTENTS

The Assessment of Science Meets the Science of Assessment

The Assessment of Science Meets the Science of Assessment

OVERVIEW

Many states and school districts are implementing reforms of science education aimed at helping all students acquire deep knowledge of scientific concepts and advanced thinking skills. Often these reforms involve new teaching methods that encourage students to explore key science topics in depth, work on realistic and complex problems, think like scientists, and integrate knowledge from several domains. In many states, science education reforms are linked to broader efforts to improve the entire educational system and to set high standards for student learning.

Assessment has become an increasingly critical part of these reforms. States and districts are revamping their assessment systems to make them more compatible with new science and math education standards and with innovative teaching methods. Interest is growing in alternative forms of assessment that ask students to demonstrate their performance by solving open-ended problems, doing projects and presentations, and collecting portfolios of their work. Researchers are working with classroom teachers to pilot techniques—such as encouraging students to assess themselves or teachers to formalize their judgments—that can make assessment a more integral part of everyday teaching and learning.

These developments in science assessment raise many technical and policy issues about the reliability, validity, fairness, and cost of new forms of assessment, and whether these new assessments are feasible and suitable for applications on a wide scale or with high stakes attached.

PURPOSE OF THE NRC CONFERENCE

To explore the connections between new approaches to science education and new developments in assessment, the Board on Testing and Assessment (BOTA) of the National Research Council (NRC) sponsored a two-day conference on February 22 and 23, 1997.

Participants included BOTA members, other measurement experts, and educators and policymakers concerned with science education reform. The conference encouraged the exchange of ideas between those with measurement expertise and those with creative approaches to instruction and assessment.

Participants presented recent research on science assessment, teaching, and learning and discussed efforts to improve assessment practice and classroom practice. They debated the role of assessment in the systemic reform of science education. The conference also featured a poster session in which participants could see firsthand some of the new science assessment tools being used in several schools.

The conference gave particular attention to these questions:

- What are the various roles that assessment is expected to play in science education?

- How well do existing assessment tools align with the kinds of scientific knowledge and skills that students should have?

- Which kinds of new assessment strategies are needed to support new instructional approaches?

- What are some innovative approaches to assessment in science education, and what effects do they have in classrooms?

- What evidence is available about the technical quality and feasibility of alternative assessments? How much do they cost?

- How well can new assessment methods meet the need for accountability in educational programs? Are they suitable for widespread use or for high-stakes situations?

- What practical and policy issues are raised by alternative approaches to science assessment? Are the alternative approaches credible and understandable to parents, the public, and policymakers? How soon can they be implemented?

This report summarizes the presentations, papers, and group discussions from the conference. In the four main sections below, the report (1) describes several alternative assessment methods presented at the conference, (2) explores major issues related to their design and use, (3) discusses how alternative assessments are being used to improve instruction, and (4) examines the implications of using various forms of assessment to promote systemic reform and accountability. A concluding section looks at avenues for meeting the complex demands for new kinds of assessments that serve instructional and accountability purposes.

PURPOSES OF SCIENCE ASSESSMENT

Assessment has multiple purposes. One purpose is to monitor educational progress or improvement. Educators, policymakers, parents, and the public want to know how much students are learning compared to standards of performance or to their peers. This purpose, often called *summative* assessment, is becoming more significant as states and school districts invest more resources in educational reform.

A second purpose is to provide teachers and students with feedback. The teachers can use the feedback to revise their classroom practices, and the students can use the feedback to monitor their own learning. This purpose, often called *formative* assessment, is also receiving greater attention with the spread of new teaching methods.

A third purpose of assessment is to drive changes in practice and policy by holding people accountable for achieving the desired reforms. This purpose, called *accountability* assessment, is very much in the forefront as states and school districts design systems that attach strong incentives and sanctions to performance on state and local assessments.

As testing experts so often stress, certain kinds of assessments are better for some purposes than for others. Good testing practice emphasizes the importance of using a test only for the purpose for which it was designed. Because good testing practice also means properly interpreting and reporting test data, the audience for the assessment data is another important consideration in test development. In addition to students and teachers, the major audiences for publication of test scores include parents, school administrators, policymakers, researchers, and the general public.

Particular care must be taken when tests are used for high-stakes decisions about people or institutions, such as student promotion, college attendance, teacher career paths, or school funding allocations. Tests used for high-stakes applications, or accountability to an external authority, or publication of scores to a wide audience must meet strict psychometric criteria (technical properties established through measurement science). For example, these tests should demonstrate adequate reliability (scores are consistent and generalizable to the broader universe of knowledge and skills) and validity (the test really measures what it is supposed to and can form a basis for appropriate inferences). They should also be fair (not biased based on factors such as ethnicity or gender).

A critical issue for the development of innovative approaches to science assessment is the usefulness of these approaches for the various purposes of testing. Standardized multiple-choice tests continue to prevail in large-scale assessment programs used for accountability and other high-stakes purposes largely because they meet the *technical* criteria of reliability, validity, and fairness and are low-cost and efficient to administer and score. But many educators and researchers question the broader validity of traditional multiple-choice tests, arguing that they do not fully capture the kinds of conceptual knowledge and thinking skills embodied in new standards that outline desirable learning goals. Because of this, existing tests might not adequately support reforms in science education and might discourage teachers from adopting effective strategies for teaching advanced knowledge and skills.

The developers of new assessment approaches are trying to overcome the limitations of multiple-choice tests by broadening the available types of assessment tasks and methods and by more directly tying them to classroom practices. But innovative assessments have their own drawbacks. They are generally more expensive and complicated to administer and score. Most have not achieved the technical quality desired for accountability and other large-scale or high-stakes purposes. Developing innovative assessments has proved to be an exceedingly difficult problem that will require expertise from both measurement science and educational practice.

INNOVATIVE FORMS OF SCIENCE ASSESSMENT

The conference highlighted a variety of efforts under way to develop and implement innovative approaches to assess science learning. The designers of these assessments are motivated by several goals. They want assessments that better measure advanced knowledge and skills, are more closely aligned with state and local standards, and will tell people how students are progressing toward standards. Many innovators are also trying to influence instruction by making assessments more compatible with new teaching strategies and by using tasks that model desirable classroom practices. Some also seek to involve students more in their own assessments—and their own learning—by giving them meaningful tasks and encouraging them to reflect on what they are doing.

Many of these alternative assessments involve open-ended items, complex problems, and performance-based tasks. Many also entail nontraditional methods of response and scoring. Several take a longer-term view of assessment, requiring students to do in-depth experiments or projects or collect examples of their work over time in a portfolio. In contrast to traditional assessment, which often takes place at the end of learning when it is too late to make a difference, many innovative assessments encourage teachers and students to engage in continuous cycles of problem solving, reflection, discussion, and revision.

PERFORMANCE ASSESSMENT

Many of the alternative approaches presented at the conference fall within the broad category of performance assessment, which refers to any assessment that requires students to demonstrate their knowledge or skills by creating an answer or product. Performance assessment includes a wide range of activities, from responding to a short-answer essay question to conducting a semester-long project.

Performance assessment aims to provide richer information about student capabilities, such as planning and conducting an experiment and working collaboratively, that are not captured by traditional test formats. It is particularly focused on assessing what one can do, as distinct from what one knows. For example, as Bert Green of Johns Hopkins University observed, someone might know how to play volleyball but no longer be able to do it physically.

The Scientific and Measurement Arenas for Refining Thinking (SMART) Project, a middle school curriculum described at the conference by Susan Goldman and James Pellegrino of Vanderbilt University, uses various kinds of performance assessment for both formative and summative ends. SMART instructional activities include working on familiar, attention-grabbing problems of water quality as a way to develop the students' deep knowledge of scientific concepts and the ability to monitor their own understanding. Students design and discuss problem-solving strategies, receive guided support from teachers, obtain feedback from classmates and Internet buddies, and eventually conduct hands-on water testing at local rivers.

As with many new curricular projects, assessment is integrated into the ongoing teaching and learning activities that make up the substance of the program. Assessment occurs through presentations, projects, discussions, and paper-and-pencil tests. For example, as a final SMART task and summative assessment, students make a group oral presentation to their classmates. Students score each other using class-developed criteria. The presentations are also videotaped, and the tapes are submitted to a central team of teachers, who review them and send some to the scientists at the state water quality department. In evaluating the presentations, the teachers and experts look at criteria such as how well students identify the main issues and how completely they understand key concepts and scientific relationships. The project also includes some performance items in its written exams used for pre- and post-testing. For example, students are asked to make a drawing explaining the concept of dissolved oxygen; those who have learned the desired concepts will produce more sophisticated drawings showing molecular structures.

Performance assessment can also be administered as part of a large-scale standardized test. The 1995 Third International Mathematics and Science Study (TIMSS), described at the conference by Maryellen Harmon of Boston College, was the largest and most technically sophisticated international achievement test ever conducted, involving half a million students in 41 countries. A much smaller sample of 6,000 students from various countries took a set of 12 performance items in science and math, each either 15 or 30 minutes long, that were designed to measure skills that could not be assessed with the multiple-choice items on the basic TIMSS test. Some tasks measured a single skill, such as using a thermometer, while others involved several steps to replicate complex real-world situations, such as determining and explaining the effect of exercise on one's own heart rate.

PORTFOLIO ASSESSMENT

Portfolio assessment measures performance by collecting portfolios of student work representing what they know and can do. One effort to grapple with the challenges of portfolio assessment was presented by Elizabeth Stage, co-director of the New Standards Project science initiative. In this project, several states and urban districts are working together to design an assessment system that will measure how well students are doing compared with the voluntary content standards emerging from national professional organizations in science, math, and other disciplines. The New Standards assessment system includes three related elements: (1) performance standards in various subjects based on the voluntary content standards; (2) external examinations that measure student performance in reference to these standards, using a combination of multiple-choice and performance tasks; and (3) a portfolio assessment system. The last two elements are still under development.

Each New Standards portfolio contains entries from an individual student. These entries include papers, experiments, projects, and commentaries on books read, which demonstrate that the student has met certain performance standards. Maintaining consistency of portfolios across students and classrooms is a particular challenge, which the New Standards Project has tried to address with specific criteria for portfolio entries. In science, the entries must illustrate the students' understanding of key concepts in physical, life, earth, and space sciences and in scientific connections and applications; entries must also demonstrate skills in scientific thinking, communication, investigation, and use of tools and technologies. For instance, a student might demonstrate her understanding of the properties of matter (a key physical sciences concept) by entering in her portfolio a write-up of a laboratory investigation of how volume and mass are related to density. This write-up might also demonstrate good communication skills by showing that the student can write clear instructions for an experiment.

To help teachers and students understand what high-quality student work looks like, the New Standards Project is collecting pieces of authentic work that exemplify high performance at different grade levels, along with explanations of how that work was evaluated. To make the point that students in all types of classroom situations can meet high standards, the project is trying to amass diverse examples of high-quality work from a variety of classrooms with different instructional programs.

The New Standards Project sees several potential benefits of portfolio assessment. It can make performance expectations more explicit and more connected to what students do in the classroom. It allows educators to assess knowledge and skills that require extended time and opportunities for revision (in contrast to knowledge and skills that students should be able to demonstrate on demand). Portfolio assessment enables teachers and students to have sustained conversations about the quality of students' work and to share responsibility for the outcome. Although optimistic about the instructional value of portfolios, Stage contends that portfolio assessment is a long way from achieving the technical quality necessary for large-scale assessment (e.g., consistency of scoring), and thus makes more sense as one of several components of an assessment system.

CONCEPT MAPS

A concept map is a graph in which nodes represent key concepts, lines linking the nodes represent relationships between these concepts, and labels on the lines describe how the concepts are related. A pair of concepts and a labeled linking line are called a proposition, the fundamental unit of a concept map. A concept map is intended to document an important structural aspect of an individual's knowledge about a content domain, for example, whether a student understands the process of photosynthesis and the relationships of carbon dioxide and oxygen to life cycles.

An accepted instructional tool, concept maps have recently become the subject of research about their usefulness as assessment tools; pertinent studies include work by conference presenters Maria Ruiz-Primo of Stanford University and Harry O'Neil of the University of Southern California. A concept map becomes an assessment tool when it is used as a format for students to give evidence of what they know about a particular domain and when it is accompanied by a scoring system.

According to the presenters, concept maps have some apparent assessment advantages. They can measure certain kinds of conceptual understanding in more depth than multiple-choice tests. They require less student writing than some other types of performance assessments, and often cost less to administer and score. Students can collaborate on constructing concept maps, which means the maps can be used to assess teamwork skills, such as leadership, decision making, and communication, as well as individual skills. A disadvantage is that concept maps can be difficult for parents and the public to understand, and thus are not as credible as more familiar kinds of assessment.

Assessments based on concept maps can vary considerably in their tasks, response modes (oral, written, or computerized), and scoring systems. Tasks can be designed to provide test takers with varying degrees of support. Some concept map assessments ask students to generate their own concepts for a particular domain, while others give students a list of concepts from which to construct their map. Still others give students a map with concepts already filled in and ask them to label the connecting lines. The research of Ruiz-Primo indicates that the first two formats—letting students generate their own concepts or giving them a list—produce equivalent student scores, both in the accuracy of students' propositions and the proportion of students' propositions that are valid. The two formats also produce similar reliability and validity coefficients. Providing students with lists of concepts makes scoring easier, however, and is therefore more feasible for large-scale assessment (although assessment designers must be careful to choose appropriate concepts). Concept maps can be scored by checking whether the map includes specific features, such as numbers of nodes and accurate propositions; by comparing the student's map with a map developed by an expert; or by combining these two strategies.

According to the presenters, research to date indicates that concept maps are a feasible form of assessment and can provide good estimates of student competence. Ruiz-Primo compared concept map scores with scores on multiple-choice tests and found that correlations between the two are moderately high ($r = .57$ on average); this suggests that these two types of assessments measure overlapping yet somewhat different aspects of knowledge. O'Neil's work with collaborative concept maps found a correlation of $r = .7$ between concept maps and an essay task.

In short, the technique is promising, but further research is needed before concept maps become a proven assessment tool. Among the issues to be addressed are: How large a sample of concepts needs to be assessed to capture a student's knowledge structure? Are different response modes interchangeable?

COGNITIVELY GUIDED ASSESSMENT

Another trend in assessment is to apply advances from cognitive science about how people think and learn to the process of designing and evaluating tests. As described by Robert Glaser of the University of Pittsburgh, the aim is to produce an assessment that effectively measures complex cognitive skills, such as the ability to integrate knowledge and make inferences, select and execute strategies for solving problems, adjust one's own performance, and offer coherent explanations for the strategies used.

Under this approach, a test developer determines the construct validity of a test by paying attention to its cognitive validity, that is, by looking at whether the test is a valid measure of both the content knowledge and the cognitive skills required by a particular domain, and whether it engages students in the kinds of cognitive activities that underlie competent performance. When a test has cognitive validity, students who score well on the test are those who demonstrate high-quality cognitive activities.

Assessment tasks can draw upon many combinations of content knowledge and process skills that place varying demands on the test taker. Robert Glaser and Gail Baxter at the University of Michigan have developed a content-process space model that can be used to analyze the cognitive demands of existing assessments. In this model, the task demands for content knowledge are placed on a scale from rich (tasks that require an in-depth understanding of subject matter) to lean (tasks that depend on information given at the time of the test rather than on prior knowledge). The task demands for process skills are placed on a scale from open (students are given a minimum of explicit direction and must decide themselves how to solve the problem) to constrained (e.g., students are given step-by-step instructions).

How a task is classified on these dimensions reveals much about the nature and extent of the cognitive activity that underlies performance. For example, a content-rich, process-constrained task might ask students to describe the possible forms of energy and types of materials involved in growing a plant, and to explain how they are related. To succeed in this task, a student would need an understanding of photosynthesis and the ability to develop a coherent explanation, but would not need to plan a solution strategy. A content-rich, process-open activity might ask students to design and carry out experiments with a maple seed to explain its flight to a friend who has not studied physics. To complete this task, students would need knowledge of force and motion, the ability to design a controlled experiment, and skills in model-based reasoning.

Test designers can adjust the content and process demands to align assessments with performance objectives. Glaser and Baxter are studying several trial state performance assessments and have conducted observations of students as they complete science tasks and solve problems in different ways and contexts. The research goal is to determine whether these tests actually measure the cognitive capabilities associated with various levels of student achievement, whether the tasks accurately reflect the performance objectives, and whether the scores reflect the quality of the cognition exhibited by test takers. The researchers have concluded that it is not easy to translate educational goals into test objectives, and then into assessment situations that maintain the integrity of those goals. Looking at the cognitive qualities of student competence, however, is a critical step in this process.

The current push to apply findings from cognitive science research—about the structure of content domains and the development of students' competence in those domains—to the development of educational assessments is an important area that was not addressed in great depth at this conference. Questions to consider include: What is the connection between theories of domain knowledge and its long-term acquisition, on one hand, and performance assessment schemes, on the other?

DESIGNING AND USING INNOVATIVE ASSESSMENTS

The conference explored several issues concerning the design and use of innovative assessments.

TECHNICAL ISSUES

A major issue in designing alternative assessments is the need for empirical evidence of reliability, validity, and fairness. Several measurement experts at the conference felt that innovative science assessments are "not ready for prime time," as Bert Green noted. These participants expressed the view that, although these assessments might be useful for improving instruction, they do not have sufficient validity and reliability to be suitable for accountability decisions and other high-stakes uses in U.S. schools today.

When these criteria are applied, certain types of performance assessment are closer to being ready for use in accountability than others. The TIMSS performance assessment tasks were developed according to technical criteria typically applied to large-scale standardized assessments. And concept maps also show promise for wider use.

Scoring Issues

The degree to which there is consistency in scoring is a primary concern related to the reliability of performance assessments. Most performance assessments must be scored by people (in contrast to assessments using multiple-choice items, which can be scored by machine). Raters of performance assessments are asked to make complex judgments, which opens the potential for variability across raters and bias in scoring. To achieve consistency, designers of performance assessments develop careful scoring criteria, provide extensive training to raters, usually use more than one rater (at least on a subset of student responses to check reliability), and develop procedures for cross-checking discrepancies.

Based on cognitive research and field trials, the TIMSS designers developed an extensive scoring system for the performance items. This scoring system was based on descriptions of typical student responses that exemplified different levels of student performance, and on common types of misconceptions that students were likely to demonstrate. One TIMSS item, for example, asked students to determine the effect of temperature on the rate of solution of an Alka Seltzer tablet and to explain their hypotheses. A student who referred to the greater energy levels of hot water molecules, and thus the faster speed with which the tablet dissolves, would get a relatively high score, and an even higher one if she adequately explained how she planned her investigation. Because test designers could not foresee all possible student answers, the scoring system also allowed raters some latitude to make judgments about responses that did not fit any of the descriptions; these judgment scores were double-checked by groups of experts in every country.

Much attention has also been devoted to systematizing the scoring of portfolio assessments. In the New Standards Project, teachers evaluate portfolios according to specific guidelines. For example, teachers look first at whether the portfolio includes all the required entries and whether each entry is an appropriate submission for a particular standard. If these basic criteria are met, the teacher then scores the quality of exhibits according to specific rubrics, such as whether the student demonstrated understanding of an important concept, represented it in multiple ways, and explained it well. Good scorers are those who understand the standards and levels of expectations and who can apply the rubrics consistently instead of using idiosyncratic processes and criteria.

Research on scoring of portfolio assessments for writing indicates that, with good training, experienced scorers, and good benchmarks, portfolio assessment can reach a level of inter-rater agreement close to the level found in writing assessments with standardized, more constrained tasks. But in science, this level of inter-rater agreement has not been achieved.

Concept maps show high reliability of scores across raters (coefficients above $r = .90$), according to Ruiz-Primo, high enough that one rater can provide a reliable score. Klein's research on hands-on assessment also found high inter-rater correlation (about $r = .95$), but these high levels of agreement do not come easily. It is necessary to provide raters with extensive training *on the specific assessment task* and with continuing supervision—a type of training that is not feasible for raters of portfolio assessments because the tasks are not the same across students.

Comparisons with Other Methods

Analyses of reliability often compare alternative assessments with multiple-choice tests in the same domain. As a general rule, reliability is higher when a test contains more items; multiple-choice tests have the general advantage, therefore, that more items can be answered in a brief amount of testing time.

Stephen Klein of the RAND Corporation presented data comparing multiple-choice tests with large-scale, hands-on performance tests. (An example of the kind of performance task used in the latter tests is to have students measure the amount of force required for different levers to lift a weight, then to have them record data, draw inferences, make predictions, and describe why a certain factor was important.) Klein's research concluded that hands-on tests are substantially less reliable per hour of testing time than multiple-choice tests. Hands-on assessment usually requires at least three to four times as much testing time to produce total scores that are as reliable (r =.9) as one period of testing with the Iowa Test of Basic Skills. If the unit of score reporting is aggregated, for example to the classroom or school level instead of the student, it takes about two periods of hands-on testing to reach this level of reliability.

Klein and others examined the correlation of hands-on assessment tasks with multiple-choice scores, and found a correlation of r =.55, the same correlation obtained between two sessions of hands-on testing. O'Neil obtained a correlation of r =.7 between concept maps and an essay-based task on the same topic; Ruiz-Primo obtained an average r of .57 between concept map scores and multiple-choice tests.

NEED FOR SCAFFOLDING

As discussed by Maryellen Harmon, some of the first science performance assessments designed for TIMSS were quite open-ended, providing little guidance to students so that they could design their own strategies and apply their learning in new ways. But many students tended to veer off on tangents, missing the main point of the task. Developers of items found that more useful information was often obtained by providing students with "scaffolding," that is, structured information, guidance, and cues that help them proceed with the task. But how much scaffolding should be provided continues to vex researchers. On one hand, students need enough scaffolding to keep them focused on the task. By narrowing the variety of student responses, scaffolding can also make scoring easier. On the other hand, too much structure can diminish the problem-solving nature of a performance task and enforce one particular approach to solving a problem. Thus, there is a need to be careful about the character of scaffolding—to ensure that it does not block multiple reasonable approaches to a task.

The designers of the TIMSS items decided to provide different levels of scaffolding for different age groups, after field trials showed some performance tasks to be too open-ended and vague for younger test takers. For example, their redesigned version of the heart-rate task now specifically directs nine-year-old children to measure their pulse for 10 seconds, enter the result in a two-column table, and identify the effects of exercise after a certain number of measurements. The 13-year-olds' version of this task leaves it up to the students to decide which measurements to make, how many to make, how long to make them, and how to record them; a final question asks them to explain why the pulse changed in the way that it did.

TECHNOLOGY

Technology is an integral part of several innovative assessments discussed at the conference. Technology can help with the design, administration, and scoring of performance assessment and can help reduce the burden on teachers when more formative assessment is added to the instructional process. O'Neil, for example, illustrated how computer software can facilitate administration, scoring, and reporting of concept map assessments, can reduce scoring costs, and can provide helpful feedback on student proficiency in specific cognitive areas. Teachers involved in Jan Hawkins's project on collegial judgment, described below, found "digital portfolios" to be a useful tool for collecting student work, particularly because the digital portfolios could be adapted to meet specific classroom needs.

Some participants, however, raised caveats about technology and assessment. For instance, research suggests that different modes of delivery—such as computer versus paper—might produce different results, even when tasks are otherwise the same.

ROLE OF READING, WRITING, AND COMMUNICATION SKILLS

Many performance-based science assessments require more reading, writing, and oral communication skills than multiple-choice tests. For students who have difficulty reading long passages, writing an explanation, or making a speech, performance tasks can be unfair if these types of extraneous task demands interfere with students' ability to demonstrate what they actually know about science. Even poor handwriting can affect scores on an open-ended question. In nearly every country, students who could solve problems correctly on the TIMSS performance tasks had considerably more difficulty trying to explain why something happened or how they figured out the answer.

Good reading, writing, and communication skills are indispensable to the practice of science, as reflected by their inclusion in voluntary science education standards. Several conference participants felt that "knowing" science includes a foundation in these skills, and that it is appropriate for science programs to teach and assess them. Nevertheless, the main purpose of science education is to build students' knowledge of scientific content and their scientific thinking skills, and effective assessments must be able to distinguish and clearly measure this knowledge and these skills. How to disentangle knowledge of scientific content and skills from reading and communication skills warrants further research.

COSTS

Performance assessments are considerably more expensive than multiple-choice tests, a major impediment to their broader use. Performance assessment takes more testing time, costs more to administer, and often requires specially trained administrators. It also requires more expensive materials. Space needs are another cost factor. Some hands-on tasks cannot be done in a standard classroom setting; for example, students might need to be physically separated so they cannot see what others are doing. Moreover, developmental costs are continual, because performance tasks generally cannot be reused from one year to another, and therefore new tasks must be developed for each test administration. Most kinds of performance items must be scored by people instead of machines, which adds substantially to scoring costs. According to Steve Klein, producing individual scores on performance assessment can cost $10 per student, compared with about 5 cents per student to score a multiple-choice test.

The experience of other countries that use large-scale performance assessments sheds light on their costs. In one example, explained by Paul Black of King's College, London, the United Kingdom has an exam for advanced-level high school students that includes essays and laboratory problems as well as multiple-choice items. This exam costs the equivalent of $43 per student per subject, or about $120 per student for the three subjects tested. The external exams taken by all 16-year-old students in the United Kingdom include written papers and some teacher-controlled assessment, at a cost of $24 per subject per student.

WEIGHING THE BENEFITS OF PERFORMANCE ASSESSMENT

Given the technical limitations and costs of performance assessments described above, why push to use them? Opinions at the conference differed as to whether the benefits of performance assessments justify their costs.

Some countries willingly accept the high costs because they believe it is important to base accountability on various measures. Advocates in the United States similarly argue that $50 per student for performance assessment is not an unreasonable investment to produce a testing system that works better. A main argument for pursuing performance assessment is its ability to encourage teachers and students to engage in more desirable classroom practices (discussed at length in the next section). Advocates contend that, in light of the added teaching and learning value of performance assessment, its costs more accurately should be compared with testing *and* curriculum development instead of just testing. It is also argued that the increased validity of performance assessment justifies its cost—if assessments are to be aligned with standards, then that may require some use of performance tasks.

Other conference participants asserted that there is still little proof that performance assessment connects to higher student learning. Because of this lack of evidence and the limited utility of performance assessment for high-stakes decisions, they find it difficult to justify its extra costs, especially if the same basic news about student achievement can be collected with less-expensive assessments.

ASSESSMENTS FOR CLASSROOM IMPROVEMENT

Many of the innovative projects discussed at the conference have the explicit goal of using assessments to improve instructional practice. Most are in the early stages of development and implementation. The basic idea behind them is that the assessment process itself can be a tool to help teachers and students improve their performance. By doing the kinds of performance-based tasks described in the previous section, students gain experience solving problems and monitoring their own understanding. Teachers gain experience in leading a classroom where students learn by inquiry and discovery.

These approaches place the formative purpose of assessment at the forefront, providing teachers and students with information they can reflect on and use to revise their own teaching and learning. As Margaret Brown of King's College, London, observed, the traditional "teach a skill then test it" approach does not fit with research showing that instructional innovations, which are designed to strengthen the frequent feedback that students receive about their learning, yield substantial gains in learning.

Some designers of innovative science learning environments have begun building assessment strategies into their projects from the very beginning. Many have the goal of embedding assessment strategies so thoroughly into everyday classroom practice that the assessments become virtually indistinguishable from other teaching and learning activities.

ENCOURAGING SELF-ASSESSMENT AND THINKING SKILLS

Several projects are aimed at developing students' abilities to monitor and reflect on their own learning and become better judges of their peers' work. An example is the

SMART Project, which engages students in repeated cycles of activity during which they make choices, reflect, discuss with others, weigh feedback, and revise their thinking. As they learn from these experiences, they proceed to more complex tasks.

The ability to integrate knowledge is another critical skill in science, enabling people to become better problem solvers and apply science in a variety of contexts throughout their lives. Marcia Linn at the University of California, Berkeley, has sought to identify a repertoire of assessments and activities that engage students in knowledge integration, which she defines as a dynamic process of linking, connecting, distinguishing, organizing, and reflecting on models of scientific phenomena. Both the curriculum and the assessments in this project encourage students to reflect on what they are learning, continuously revisit ideas, critique the ideas of others, and refine their understanding—similar to the way in which professional scientists regularly reformulate and improve scientific ideas.

As an example, one area that Linn has been working on is helping students to develop an integrated understanding of heat and temperature, since she has found that people have a multitude of models, based on their everyday experiences, about the distinction between these two concepts. One assessment-related activity in her repertoire requires students to predict whether various objects would be above, below, or at room temperature after being placed for eight hours in the same room. After instruction in thermal equilibrium, students answer the same questions and are then interviewed by the teacher about how and why their understanding changed from their initial predictions. Teachers and researchers have used students' responses to this task to examine the development of conceptual understanding in this domain and to make successive refinements in the curriculum.

DOCUMENTING CLASSROOM LEARNING

Much of the evidence that guides teachers' daily decisions about instruction comes from their own informal observations of children's behavior—forms of evidence that are often sufficient for teachers' ongoing decision making but tend to be difficult to document. For situations in which more precision is required, Edward Chittenden of the Educational Testing Service described two different projects in which teams of teachers have developed strategies for documenting what children are learning in science by examining the many ways in which the children demonstrate their understanding through talk, work products, and responses to questions and quizzes. These projects seek to develop teachers' abilities to observe, listen, record, and take a second look while they build documentation of learning over time and across settings. Teachers can then use this information to plan more effective instruction. A long-term goal of the projects is to develop documentation methods that could be informative to parents, administrators, and others outside the classroom and perhaps could be aggregated and summarized.

In one project, kindergarten teachers wrote down key things that children said about their drawings of light and shadow. In another, teachers charted changes in children's answers to such questions as, "What are some things you've noticed lately about the caterpillars?" Through careful observation, teachers can identify what children already know, what misconceptions they have, and where they have gaps in understanding. Teachers reported being surprised at what students who are lagging academically actually know.

ASSESSMENT AS PROFESSIONAL DEVELOPMENT

These new forms of assessment depend heavily on implementation by teachers. Many teachers have not been trained to use assessment in the ways envisioned by science education reforms, and few have in-depth knowledge of evaluation issues. Margaret Brown found that more than half of the teachers studied emphasized summative assessment over formative assessment; indeed, many considered formative assessment an unnecessary addition to their workload. These teachers taught according to the prescribed curriculum and did not adjust their teaching based on assessment-related feedback.

But another group of teachers—about one-third of those studied—did conduct regular formative assessment, even when it was not required. Some of these teachers relied on observation, note taking, and discussion, while others organized distinct assessments. All used the feedback they collected within a short time to plan classroom work for subsequent weeks. The top teachers in the study, as gauged by student gains, were focused on assessment all the time.

Several projects discussed at the conference use assessment expressly as a vehicle for the professional development of teachers. Margaret Brown found that, with the right kind of professional development, it is possible to encourage teachers to use formative assessment more often and more effectively. Teachers are trained to assess at the beginning, middle, and end of a topic. Over time, they internalize the assessment criteria and integrate assessment more into their teaching. Teachers develop enthusiasm for this kind of assessment, as long as it is not being used for high-stakes purposes.

Many of the new science instruction methods also require teachers to interact differently with students and manage a more open classroom. But many teachers do not have the experience to lead this kind of instruction, nor a clear vision of what it should look like. In addition, to teach to high standards, many teachers will need a better mastery of content knowledge. As Joan Baron of the Connecticut Department of Education noted, "If we want students to internalize standards, then teachers must be comfortable with them." Several projects presented at the conference are attempting to use assessment not only to build teachers' evaluation skills, but also to develop their pedagogical skills and deepen their understanding of the content they are teaching.

Collegial Judgment of Student Work

The move to embed assessment in classroom practice has increased teachers' needs for more formal and consistent strategies for judging student work and using those judgments to improve instruction. As explained by Baron, the Connecticut Department of Education has used Goals 2000 funds to encourage teachers to develop a process for holding frequent and sustained conversations about students' work. In one project, teachers looked at student work alone, and then discussed it in groups, identifying the main mathematical ideas they were trying to teach and the major areas where students were having problems. After reviewing promising curricula and scoring techniques, the teachers worked in pairs to develop activities to address areas of student difficulty. The teachers taught their units, making revisions as needed. As students revisited tasks that had given them trouble earlier, teachers tracked the work of three students at different achievement levels and periodically reviewed it in group meetings as a means of documenting changes in their teaching.

In this project, university and state education department researchers studied the consequences of the participating teachers' conversations about student work. They concluded that teachers needed to be given a specific set of criteria to keep them focused on the students' understanding of mathematical ideas, rather than simply telling stories about what the students did. The researchers also found that soliciting external feedback from teachers in other districts was an effective motivator and kept conversations from becoming too ingrown. Furthermore, they concluded that combining teachers from different grade levels encouraged teachers to explore the vertical articulation of curriculum and helped them identify incorrect ideas that students were retaining over time.

Jan Hawkins of the Center for Children and Technology described the Center's projects to help teachers learn to make reliable and useful judgments about student work. Teachers in one project judged three kinds of student performance tasks: a final product, an oral explanation, and a process record (e.g., journal or log). Originally the project gave teachers a set of rubrics for judging student work, but the researchers realized a more effective approach was to provide teachers with a wide variety of student examples and allow them to arrive at their own rubrics through collective deliberation and experimentation. Over time, the teachers designed their own guidelines for making judgments, and they learned how to moderate scores collegially and cite evidence for their judgments.

Hawkins has identified conditions that seem to affect consistency in evaluations by teachers. First, teachers' judgments are better, more consistent, and more reliable when teachers cite specific evidence for them. Second, the moderation process by which several teachers reach a uniform judgment—especially its public nature and social dimension—is very important and cannot be omitted. Third, the type of task being judged is a critical factor; some tasks are more amenable to collaborative judgment than others. Fourth, individuals who have more content expertise tend to be more consistent with each other, and can be helpful when they are paired with someone with less expertise.

ISSUES IN CLASSROOM-BASED ASSESSMENT

Discussion at the conference raised some critical issues related to using classroom formative assessments for improving instruction.

Effectiveness in Changing Instruction

To what extent are these assessment approaches actually making a difference in the classroom? According to conference presenters, teacher-based assessment projects help teachers become more systematic in collecting their observations and more reliable in judging student work. By seeing and discussing exemplary work, they begin to internalize standards and adjust instruction based on their assessments. When teachers achieve these goals, students benefit because they receive more regular and useful feedback, as well as models for self-assessment.

Jan Hawkins said that her project on collegial judgment did improve teaching, but slowly. Rather than adopting the rubrics that came out of the project, participating teachers used their new knowledge to redesign their classroom activities and modify their judgments about student projects. It was critical to involve students in this process, too, so they understood the kinds of judgments being brought to their work.

Some participants felt that the effects of formative assessments in changing classroom practice tend to be overstated. Others believed that new assessment methods would be more likely to drive change if the results had consequences and if they could be incorporated into accountability systems, as discussed below.

Scaffolding for Teachers

The projects discussed in this section suggest that teachers need appropriate scaffolding, too. This means providing teachers with a great deal of support and guidance during the early stages, as they are learning to use new forms of assessment, followed by diminishing support as teachers gain competence in the new techniques. Several projects achieved this by providing teachers with some type of framework or information for analyzing assessment data. Having a framework encourages teachers to work systematically rather than intuitively, and it also raises teacher professionalism by exposing them to new techniques. But the framework should not be too prescriptive; as Hawkin's experience indicates, teachers learned a great deal by working through the process themselves.

Other Teacher Supports

Participants noted that teachers who are implementing new approaches to assessment and instruction need other forms of support and capacity building to help their efforts succeed. First, teachers need time in the school day for collegial discussions and other activities related to formative assessment. Second, a school should have a learning culture that encourages teachers to work together and supports new modes of student-teacher interactions. Third, assessment alone is not enough to help teachers gain the knowledge and skills necessary to teach to high standards. There is a critical need for professional development in content and pedagogy at all stages of a teacher's preparation and career. This is a major undertaking that will require political support and significant resources.

ASSESSMENTS FOR ACCOUNTABILITY AND SYSTEMIC REFORM

Standards-based reforms place increasing demands on assessment. Policymakers and educators expect assessments to signal—and if possible clarify—what is important for teachers to teach and students to learn. Reformers expect assessments both to drive changes in curricula and instruction and to measure the effects of those changes. Assessment is further intended to form the backbone of new state and local accountability systems. Conference participants discussed whether and how these expectations could be met with various forms of assessment.

ALIGNING ASSESSMENTS WITH STANDARDS

A primary motive for revising assessment is to improve the alignment between what schools and teachers are held accountable for and what students are supposed to learn. Developing assessments aligned with standards is a daunting task. The ambiguity, vagueness, and wide variation in standards make it difficult to construct compatible assessments. As noted by Eva Baker of the University of California, Los Angeles, the standards being adopted by states tend to describe broad areas of content emphasis and general cognitive demands, so that an array of performance tasks can be inferred from any given standard. Most state standards are not in a form that provides sufficient focus or substance to make the connections from standards to assessment to everyday instruction, yet such connections are needed to provide a solid basis for evaluating the effects of standards-based reforms.

The diffusion of education authority and political power in the United States, among 50 states and thousands of school districts, makes it still more difficult to align standards and assessments and negotiate systemic reform, as pointed out by William Clune of the University of Wisconsin. This diffusion contributes to the dominance of standardized multiple-choice tests, because they are a generic instrument that states and localities can use to test the wide range of curricula.

Audrey Champagne of the State University of New York, Albany, described the challenge of writing a practically oriented performance task to assess a state standard for eighth graders. This particular standard called on students to generate mathematical models of actual situations and apply them to other situations. Champagne's group devised a task that asked students to calculate the minimum amount of a solvent necessary to clean a paint brush, then write a memo applying their findings to other industrial situations. The problem could be solved using sixth-grade arithmetic, but it also required students to integrate knowledge from several disciplines, extract key information from a written passage, organize a fairly complex task, and communicate how they reached their solutions.

A review committee agreed that the task corresponded well to the standard. But when Champagne and other researchers piloted the task with a group of college students, they discovered that many students did not really understand its point. Several complained about the amount of reading involved, and many did not know the basic facts needed to solve it, such as the number of ounces in a quart. Some students used creative responses to avoid addressing the mathematics. The reactions of the college students show the challenges involved in designing practical tasks with literacy demands that test math and science, and the complexities of scoring diverse student responses. It also gives a sense of the job teachers face in preparing students to accomplish this kind of performance task.

DESIGNING ASSESSMENTS FOR ACCOUNTABILITY

In aligning assessments with standards, many states and local districts hope to produce assessments that can be used for various external and accountability purposes, such as reporting progress to the public or attaching consequences to the performance of classrooms, schools, and perhaps individuals. Tests for accountability must meet relatively high technical standards and be feasible to administer on a large scale. They must also meet practical concerns, such as being understandable and credible to parents, policymakers, and the public. And they should encourage students to learn things that are important and useful. Tension between the uses of assessments for accountability and for instructional improvement was a major topic of discussion during the conference.

Experience of Other Countries

Paul Black examined how other countries try to resolve the tension between accountability and instructional improvement. The resolutions these countries reach often depend on how much their society trusts teachers' assessment of their own pupils. In countries in which teachers have high social status, assessments are more diverse, and teachers play a central role in all functions of assessment. Countries in which teachers are less trusted tend to place more emphasis on external control and high-stakes testing, with less latitude for teachers to make their own decisions. Germany and, to a lesser extent, Sweden exemplify the first view; England and some aspects of the system in France illustrate the second.

Countries differ in their use of teacher conducted assessments for accountability purposes. Some countries use teacher assessment as the sole source of accountability. In some of these, there might be calibration of teachers' results at the national level, or some external check on them. Other countries combine teacher-conducted assessment with external tests to produce a single result, or use both kinds of assessments but not in a combined way. Finally, in other countries, including the United States, teacher-conducted assessments are not generally used for accountability, except when they help pupils learn in preparation for external tests.

Not all assessments for accountability in other countries have high reliability. In England, for example, the reliability of some national tests is not very high. Only in the United States are assessments considered unsuitable for accountability if they are not as reliable as multiple-choice tests. According to Black, American policymakers and the public tend to undervalue what teachers do and overvalue external standardized tests. Other factors in the United States, such as college admission policies, also drive accountability assessments toward traditional test formats and away from certain kinds of alternative assessments.

Classroom Assessments and Accountability Testing

Where should U.S. policy be headed? Some conference participants expressed caution about the use of performance and classroom-based assessments for accountability and external purposes, noting that most did not have the requisite features or technical properties. For example, some performance assessments do not yield a score that can be aggregated. Some participants made the point that formative assessments are designed primarily to assess what is taught in the classroom and inform revisions in teaching, and are less concerned with testing a child's overall knowledge and skills. For purposes of certification, college admission, and other uses, the object of assessment is to find out what students actually know, regardless of what they were taught, and to measure their knowledge rather than the quality of instruction.

Other participants felt that collaboration between psychometricians and educators could lead to assessment methods that would legitimately serve both purposes and give teachers' contributions a greater role in accountability. As Clune observed, teachers already gather considerable performance information, and teacher-conducted assessment constitutes a large, ongoing testing system. Yet there is not much empirical information about the relationship of teacher-conducted tests to external exams. Nor is there well-refined information about the strengths and weaknesses of various types of assessments that would enable reformers to make intelligent choices. Instead the debate deteriorates to one of "good" versus "bad" tests, or "old fashioned" versus "new" tests. It would also be useful, Clune observed, to have more empirical evidence about the conditions under which high-stakes testing produces negative educational consequences—such as teachers' focusing instruction only on topics that are on the test, or districts' excluding certain children from testing—and whether performance assessment can help to reduce some of these problems.

Several participants suggested that performance-based assessments could be combined with multiple-choice items and other formats to produce a balanced assessment program. Various kinds of assessments could be given throughout the year, rather than one time only, to achieve a better balance.

ISSUES IN ASSESSMENTS FOR ACCOUNTABILITY

Developing new approaches to assessments for accountability requires attention to practical and political implications, as well as psychometric ones. As Ed Reidy of the Kentucky Department of Education noted, Americans often try to turn value-laden issues into technical issues to be resolved by technical experts, instead of debating them in a public forum with advice from the research, education, and policymaking communities.

Defining Performance Thresholds

A critical unresolved issue in designing assessments for accountability is: What level of performance is adequate? The passing scores have to be set high enough to stimulate students and teachers to improve, but not so high as to discourage students who initially fail to reach them from pursuing science and math courses. Nor do we know whether or how various performance increments are linked to meaningful social outcomes, such as success in the labor market.

Curriculum and Accountability

Another unresolved issue is how to allow for differences in a curriculum, such as a vocational education curriculum, within a common accountability system, and whether we should even have a common system. Many science and math education reformers emphasize advanced academic courses for all students, although an applied math and science curriculum might work better for some students. Can a limited core of knowledge and skills be defined that would allow a student to have some degree of deep understanding and scientific literacy without covering a large number of topics or advancing very far into formal training?

Student Motivation

Conference participants held diverse views on the relevance of student motivation to various kinds of assessments. In general, students are more motivated to do well on a test when it "counts," but what does that really mean? Consequences that seem to matter to students are promotion, graduation, and college acceptance. Tests also have a major impact when they are part of a student's grade. But are there other motivators, such as participation in team projects?

Students might be more motivated to learn when they have some personal responsibility for their own assessment, which is what several alternative assessments are trying to generate. President of the National Academy of Sciences, Bruce Alberts, gave an example from a medical school that introduced essay questions into major examinations that had previously been multiple-choice. The school found that students made a greater effort to learn for understanding, instead of for short-term retention.

Many current approaches to standards-based reform place the primary locus of accountability on the school or classroom. In these systems, it is the teacher or principal who is primarily held responsible for student performance; the motivating consequences are not as directly tied to the student. Diverse views were expressed at the conference about whether or not accountability systems should also include more meaningful consequences for students in order to be effective and have a real impact.

Public Engagement

Large-scale assessments, especially those used for accountability, must be credible, meaningful, and understandable to parents, policymakers, and the public. In many states and school districts, debates over assessment have been subject to increased scrutiny and involvement of policymakers and the public. Although an open process is important, it sometimes has made these debates more complex and politically charged, allowing researchers and assessment designers "few protected enclaves in which to try and fail," as Eva Baker noted.

Multiple-choice tests are familiar and comfortable to the public. Educators report being frustrated in their efforts to improve science instruction and assessment by parents who might appreciate their children's having new learning opportunities, but still prefer them to be prepared for traditional tests. As Bruce Alberts observed, if forced to choose between their child's learning something in high school and getting into a good college, some parents would prefer them to get into a good college and learn later. Other participants felt that what parents really care about is what their children know and can do, not which instructional method gets them there.

The public wants some fairly simple things, said Baker: tests that provide a quick summary of individual student performance and progress, and that provide a basis for comparisons among students and programs. The experts' admonitions that not all tests can be used for all purposes does not make sense to many policymakers and parents. To them, a good test serves multiple purposes, and if such a test does not exist, it should be invented. But this desire may be unrealistic; some purposes of assessment might not be reconcilable in a single instrument.

Educators and test designers have not developed a strategy for educating the public about assessment issues. Convincing parents, policymakers, and the public that unfamiliar forms of tests are important and useful to a student's future will require a major public engagement effort. Educators must be prepared to explain to the public what student scores on performance assessments mean. For instance, performance assessment results are typically reported in relation to public goals or standards, rather than in terms of the normative comparisons that the public is more accustomed to, so new reporting systems need to be designed that make sense to the typical parent. Support for standards-based reforms could well disappear if new assessments do not yield achievement information that parents and the public can understand and find useful.

Timetable for Change

Education is under pressure to show positive results soon after reforms are agreed to—often before they are really in place. Right now the education and research communities are spreading their energies across an astonishing number and variety of assessment-related projects. With so many reform efforts going on at once, it will be difficult to produce adequate empirical data in a timely way, and it may be impossible to convey to the public any sense of progress.

Educators and test designers do not have to immediately reinvent every component of an assessment and accountability system. The deadlines for action suggest a need for models that combine existing and new assessments. Reformers can begin by using assessments that are available now, and phase in changes by grade or subject area. Such a phased-in approach would demonstrate the cost and effectiveness of a new system. Reform could focus on accomplishing a manageable number of changes, rather than diffusing efforts across too many priorities.

CONCLUSION

The exciting developments in the field of science assessment represent a golden opportunity to better integrate assessment with curriculum and to improve teaching and learning. Paul Black echoed the view of many conference participants when he expressed a desire for stronger bridges among instructional innovators, psychometricians, and classroom teachers. Instructional innovators could contribute a vision for integrating assessment and instruction with some high-quality ideas for how to get there. Psychometricians could seek ways to make classroom-oriented assessments more compatible with accountability needs. Teachers could play a stronger role in assessment development to ensure that new approaches are credible and feasible. All three groups must communicate with the larger polity of parents, policymakers, and the public to build support for changes that will have a lasting positive impact on classroom practice.

APPENDIX A
CONFERENCE AGENDA AND PARTICIPANTS

Day 1—Saturday, February 22, 1997

8:30 - 9:00 a.m. **Welcome and Opening Remarks**

Daryl Chubin, National Science Foundation, and Office of Science and Technology Policy

Paul Black and Richard Shavelson, Conference Co-Chairs, Board on Testing and Assessment

9:00 - 10:45 a.m. **Science Learning and Assessment**

Presenters: ***Audrey Champagne***, State University of New York, Albany

Robert Glaser, Learning Research and Development Center, University of Pittsburgh

Marcia Linn, University of California, Berkeley

James Pellegrino, Peabody College at Vanderbilt University

Discussant: ***H.D. Hoover***, University of Iowa

Moderator: ***Richard Shavelson***, Board on Testing and Assessment and Stanford University

10:45 a.m. *Break*

11:00 a.m. Poster Session, Great Hall

12:00 p.m. *Lunch (poster session continued)*

1:00 - 3:15 p.m. **Innovative Approaches to Assessment**

Presenters:

Performance Assessment

Susan Goldman, Peabody College at Vanderbilt University

Maryellen Harmon, Center for the Study of Testing, Evaluation, and Educational Policy, Boston College

	Portfolio Assessment	
		Elizabeth Stage, New Standards Project, University of California
	Concept Maps	
		Maria Ruiz-Primo, Stanford University
		Harry O'Neil, University of Southern California
	Discussants:	***Margaret Cozzens***, National Science Foundation
		Bert Green, Johns Hopkins University
	Moderator:	***Jack Whalen***, Board on Testing and Assessment and Institute for Research on Learning
3:15 p.m.	*Break*	
3:30 - 5:00 p.m.	**Embedded and Formative Assessment and Classroom Practice**	
	Presenters:	***Margaret Brown***, King's College, London
		Edward Chittenden, Educational Testing Service
		Jan Hawkins, Center for Children and Technology, Education Development Center
	Discussant:	***Paul Black***, King's College, London, and Board on Testing and Assessment
	Moderator:	***Larry Suter***, National Science Foundation

Day 2—Sunday, February 23, 1997

8:30 - 10:00 a.m.	**Systemic Issues: Cost and Implementation**	
	Presenters:	***Eva Baker***, National Center for Research on Evaluation, Standards, and Student Testing, University of California, Los Angeles
		Stephen Klein, The RAND Corporation
		Ed Reidy, Kentucky Department of Education
	Discussant:	***Michael Feuer***, Board on Testing and Assessment, National Research Council
	Moderator:	***Margaret Cozzens***, National Science Foundation
10:00 a.m.	*Break*	

10:15 - 11:45 a.m. **Systemic Issues: Reform and Policy Initiatives**

Presenters: ***Joan Baron,*** Connecticut State Department of Education

Paul Black, King's College, London, and Board on Testing and Assessment

William Clune, University of Wisconsin

Discussant: ***Senta Raizen***, The National Center for Improving Science Education

Moderator: ***Janet Hansen***, Board on International Comparative Studies in Education, National Research Council

11:45 a.m.- 12:30 p.m. **Closing and Synthesis of Conference Themes**

Bruce Alberts, President, National Academy of Sciences, and Chair, National Research Council

Paul Black and ***Richard Shavelson,*** Board on Testing and Assessment

CONFERENCE PARTICIPANTS

BRUCE ALBERTS, National Academy of Sciences and National Research Council, Washington, D.C.

RUTH ANDRIONE, Baltimore City Public Schools, Maryland

EVA BAKER, National Center for Research on Evaluation, Standards, and Student Testing, University of California, Los Angeles

JOAN BARON, Connecticut State Department of Education

EDITH BEATTY, Continuous Assessment in Science Project, Trinity College, Burlington, Vermont

JERRY BELL, ECA/Sonic Images Productions, Inc., Washington, D.C.

LESLIE BENTON, National Science Resources Center, Smithsonian Institution, Washington, D.C.

HOWARD BERGER, NYC Community School District 3, New York

WENDY BINDER, National Science Resources Center, Smithsonian Institution, Washington, D.C.

EVE BITHER, U.S. Department of Education, Washington, D.C.

BARBARA BLACK, Cambridge Public Schools, Massachusetts

PAUL BLACK, School of Education, Kings College, London

MARY ANN BREARTON, American Association for the Advancement of Science, Washington, D.C.

RHONDA BROOKS, Albuquerque Academy, New Mexico

HELEN BROUGHTON, Memphis City Schools, Tennessee

KATHY BROWN, Massachusetts Department of Education

MARGARET BROWN, King's College, London

JOAN BUCKLEY, American Federation of Teachers, Washington, D.C.

JIM BYRNES, Department of Human Development, University of Maryland, College Park

ARTHUR CAMINS, New York City Urban Systemic Initiative, New York

MAURA CARLSON, Continuous Assessment in Science Project, Trinity College, Burlington, Vermont

JEANNE CENTURY, Education Development Center, Newton, Massachusetts

CATHY CHAMBERLAIN, Curriculum Development and Evaluation Department, Buffalo Public Schools, New York

AUDREY CHAMPAGNE, State University of New York, Albany

CYNTHIA CHIDEYA, Baltimore City Public Schools, Maryland

EDWARD CHITTENDEN, Educational Testing Service, Princeton, New Jersey

DARYL CHUBIN, National Science Foundation and Office of Science and Technology Policy, Arlington, Virginia

INES CIFUENTES, Carnegie Institution of Washington, D.C.

D. JOSEPH CLARK, Videodiscovery, Inc., Seattle, Washington

WILLIAM CLUNE, University of Wisconsin, Madison

ROSALEA COURTNEY, Educational Testing Service, Princeton, New Jersey

MARGARET COZZENS, National Science Foundation, Arlington, Virginia

PEGGY CROWELL, Jackson Public Schools, Jackson, Mississippi

SARAH DAVIS, Education Development Center, Newton, Massachusetts

STACY DOUGLAS, NYC Community School District 6, New York

MIKE DRYDEN, Dallas Public Schools, Texas

SHEILA DUNSTON, SMART Process, New York, New York

HUBERT M. DYASI, Workshop Center, City College of New York

REBECCA DYASI, Workshop Center, City College of New York

JAMES ELLIS, National Science Foundation, Arlington, Virginia

JANET ENGLISH, Serrano Intermediate School, Lake Forest, California

JOYCE EVANS, National Science Foundation, Arlington, Virginia

MICHAEL FEUER, National Research Council, Washington, D.C.

RAYMOND FIELDS, National Assessment Governing Board, Washington, D.C.

PAT FITZSIMMONS, Continuous Assessment in Science Project, Trinity College, Burlington, Vermont

JOY FRECHTLING, WESTAT, Rockville, Maryland

JOHN FREDERIKSEN, Educational Testing Service and University of California, Berkeley

ROBERT GLASER, Learning Research and Development Center, University of Pittsburgh, Pennsylvania

THOMAS K. GLENNAN, RAND, Washington, D.C.

DEE GOLDBERG, Spring Branch Independent School District, Spring Branch, Texas

SUSAN GOLDMAN, Peabody College at Vanderbilt University, Nashville, Tennessee

BERT GREEN, Johns Hopkins University, Baltimore, Maryland

STEVEN GREEN, San Francisco Unified School District, California

JAY HACKETT, National Research Council, Washington, D.C.

JAMES HAMOS, University of Massachusetts Medical Center, Worcester

JANET HANSEN, National Research Council, Washington, D.C.

MARYELLEN HARMON, Center for the Study of Testing, Evaluation, and Educational Policy, Boston College

QUINCY HATHOM, Memphis City Schools, Tennessee

JAN HAWKINS, Center for Children and Technology, Education Development Center, New York

HENRY HEIKKINEN, University of Northern Colorado, MAST Center, Department of Chemistry and Biochemistry, Greeley, Colorado

MARSELLE HEYWOOD, NYC Community School District 6, New York

H.D. HOOVER, University of Iowa

GREGG HUMPHREY, Continuous Assessment in Science Project, Trinity College, Burlington, Vermont

CHARLES JAMES, Carnegie Institution of Washington, D.C.

LIZ JONES, South Carolina Department of Education

MARGARET JORGENSEN, Educational Testing Service, Princeton, New Jersey

JANE BUTLER KAHLE, Miami University, Oxford, Ohio

ALLISON KAYE, Council of Chief State School Officers, Washington, D.C.

THOMAS KELLER, Maine Department of Education

STEPHEN KLEIN, RAND Corporation, Santa Monica, California

GERALD KULM, American Association for the Advancement of Science, Washington, D.C.

MORRIS LAI, University of Hawaii, Manoa

OKHEE LEE, School of Education, University of Miami, Florida

MARCIA LINN, University of California, Berkeley

KATHLEEN LUNDGREN, Minnesota Department of Education

BEVERLY T. LYNDS, University Corporation for Atmospheric Research, Boulder, Colorado

CHRISTOPHER LYON, Washington, D.C.

BARBARA MALCOLM, Buffalo Public Schools, New York

WAYNE MARTIN, Council of Chief State School Officers, Washington, D.C.

CHRISTINE MASSEY, Institute for Research in Cognitive Science, University of Pennsylvania

TOM McINTOSH, Milwaukee Public Schools, Wisconsin

SALLY MIDDLEBROOKS, Association of Science Technology Centers, Washington, D.C.

MARY ANN MILAK, Buffalo Public Schools, New York

INA MOGENSEN, Paradise Elementary School, Las Vegas, Nevada

DIANA MONTGOMERY, Horizon Research, Inc., Chapel Hill, North Carolina

PATRICIA MORSE, National Science Foundation, Arlington, Virginia

PATRICIA MOSTUE, Worcester Public Schools, Worcester, Massachusetts

CORNELIA MUNROE, American Council on Education, GED Testing Service, Washington, D.C.

DONNAJEAN NEMETH, Cleveland City Schools, Ohio

HARRY O'NEIL, University of Southern California

MICHAEL OSBOME, Memphis City Schools, Tennessee

KATHERINE PAGET, TERC, Cambridge, Massachusetts

JAMES PELLEGRINO, Peabody College at Vanderbilt University, Nashville, Tennessee

DIANE PUFF, Springfield Public Schools, Springfield, Massachusetts

EDYS QUELLMALZ, SRI International, Menlo Park, California

SENTA RAIZEN, National Center for Improving Science Education, Washington, D.C.

LILLIAN RAMOS, NYC Community School District 6, New York

LARRY READ, Polaris Project, Bartlett High School, Anchorage, Alaska

ED REIDY, Kentucky Department of Education

MARTHA ROBERTS, Jackson Public Schools, Jackson, Mississippi

THOMAS ROMBERG, National Center for Research in Mathematical Sciences Education, Madison, Wisconsin

JILL ROSENBLUM, Maine Math and Science Alliance, Augusta, Maine

ZIPORA ROTH, Institute for Research in Cognitive Science, University of Pennsylvania

WILLILEXIA ROYAL, Baltimore City Public Schools, Maryland

MARIA RUIZ-PRIMO, Stanford University

NORA SABELLI, National Science Foundation, Arlington, Virginia

PAUL SANDIFER, American College Testing, Iowa City, Iowa

JUDITH SANDLER, Education Development Center, Inc., Newton, Massachusetts

TED SCHULTZ, National Research Council, Washington, D.C.

RAMSAY SELDEN, Education Statistical Service Institute, Washington, D.C.

RICHARD SHAVELSON, Stanford University

LYNNE DEE SHAY, Paradise Elementary School, Las Vegas, Nevada

ANNABELLE SHRIEVE, Education Development Center, Newton, Massachusetts

JAMES A. SHYMANSKY, University of Iowa

SUSAN SNYDER, National Science Foundation, Arlington, Virginia

LAURIE SOMERS, Council of Chief State School Officers, Washington, D.C.

DIANE SPRESSER, National Science Foundation, Arlington, Virginia

NEVZER STACEY, Office of Educational Research and Improvement, U.S. Department of Education, Washington, D.C.

ELIZABETH STAGE, New Standards Project, University of California, Oakland

CAROLE STEARNS, Merck Institute for Science Education, Rahway, New Jersey

JOHANNA STRANGE, Fayette County Public Schools, Lexington, Kentucky

MARILYN J. SUITER, Education and Human Resources, American Geological Institute, Alexandria, Virginia

LARRY SUTER, National Science Foundation, Arlington, Virginia

MARY JANE SYREK, Dr. Charles R. Drew Science Magnet School, Buffalo, New York

SHAUN TAYLOR, Videodiscovery, Inc., Seattle, Washington

DAVID TAYLOR, Fayette County Public Schools, Lexington, Kentucky

DOMENIC THOMPSON, Baltimore City Public Schools, Maryland

JEAN TUSHIE, Minnesota Department of Education

GINNY VAN HORNE, American Association for the Advancement of Science, Washington, D.C.

JEAN VANSKI, National Science Foundation, Arlington, Virginia

EMMA WALTON, National Science Foundation, Arlington, Virginia

JACK WHALEN, Xerox Palo Alto Research Center and Institute for Research on Learning, Palo Alto, California

MARIE WICKS, Baltimore City Public Schools, Maryland

KAREN WORTH, Education Development Center, Newton, Massachusetts

LINDA WURZBACH, Council of Chief State School Officers, Washington, D.C.

APPENDIX B
PUBLICATIONS OF THE BOARD ON TESTING AND ASSESSMENT

Anticipating Goals 2000: Standards, Assessment, and Public Policy. Summary of a Workshop. Michael J. Feuer and Nancy Kober, Editors. 1995.

Evaluation of the U.S. Employment Service Workplan for the GATB Improvement Project. 1995.

A Valedictory: Reflections on 60 Years in Educational Testing. Lee J. Cronbach. 1995.

Letter Report: Evaluation of "Redesigning the National Assessment of Educational Progress." Committee on Evaluation of National and State Assessments of Educational Progress. 1996.

Letter Report: "Investigative Guidance on Fairness in Testing." Richard Shavelson, Chair, Board on Testing and Assessment. 1996.

Title I Testing and Assessment: Challenging Standards for Disadvantaged Children. Summary of a Workshop. Michael J. Feuer and Nancy Kober, Editors. 1996.

Transitions in Work and Learning: Implications for Assessment. Summary of a Conference. Alan M. Lesgold, Michael J. Feuer, and Allison M. Black, Editors. 1996.

The Use of IQ Tests in Special Education Decision Making and Planning. Summary of Two Workshops. Patricia Morison, Sheldon H. White, and Michael J. Feuer, Editors. 1996.

Educating One and All: Students with Disabilities and Standards-Based Reform, Committee on Goals 2000 and the Inclusion of Students with Disabilities. Lorraine M. McDonnell, Margaret J. McLaughlin, and Patricia Morison, Editors. 1997.

Learning from TIMSS: Results of the Third International Mathematics and Science Study. Summary of a Workshop. Board on International Comparative Studies in Education. Alexandra Beatty, Editor. 1997.

Taking Stock: What Have We Learned About Making Education Standards Internationally Competitive? Summary of a Workshop. Board on International Comparative Studies in Education. Alexandra Beatty, Editor. 1997.

Transitions in Work and Learning: Implications for Assessment. Papers and Proceedings. Alan Lesgold, Michael J. Feuer, and Allison M. Black, Editors. 1997.

Equivalency and Linkage of Educational Tests: Interim Report, Committee on Equivalency and Linkage of Educational Tests. Michael J. Feuer, Paul W. Holland, Meryl Bertenthal, F. Cadelle Hemphill, and Bert F. Green, Editors. 1998.

Letter to Secretary Richard Riley, U.S. Department of Education, and Mark D. Musick, National Assessment Governing Board. Robert M. Hauser and Lauress L. Wise, co-Principal Investigators on Evaluation of the Voluntary National Tests. 1998.

Evaluation of the Voluntary National Tests: Phase 1 Report. Lauress L. Wise, Robert M. Hauser, Karen J. Mitchell, and Michael J. Feuer. 1999.

Grading the Nation's Report Card: Evaluating NAEP and Transforming the Assessment of Educational Progress, Committee on the Evaluation of National and State Assessments of Educational Progress. James W. Pellegrino, Lee R. Jones, and Karen J. Mitchell, Editors. 1999.

High Stakes: Testing for Tracking, Promotion, and Graduation, Committee on Appropriate Test Use. Jay P. Heubert and Robert M. Hauser, Editors. 1999.

Next Steps for TIMSS: Directions for Secondary Analysis. Board on International Comparative Studies in Education. Alexandra Beatty, Lynn W. Paine, and Francisco O. Ramirez, Editors. 1999.

Uncommon Measures: Equivalence and Linkage Among Educational Tests, Committee on Equivalency and Linkage of Educational Tests. Michael J. Feuer, Paul W. Holland, Meryl Bertenthal, F. Cadelle Hemphill, and Bert F. Green, Editors. 1999.